Understanding the Elements of the Periodic Table™

HELIUM

Heather Hasan

2   4

He

The Rosen Publishing Group, Inc., New York

To my niece, Meagan, a lover of balloons. You are so precious to me.

Published in 2007 by The Rosen Publishing Group, Inc.
29 East 21st Street, New York, NY 10010

First Edition

Library of Congress Cataloging-in-Publication Data

Hasan, Heather.
Helium / Heather Hasan.—1st ed.
 p. cm.—(Understanding the elements of the periodic table)
Includes bibliographical references and index.
ISBN 1-4042-0703-1 (lib. bdg.)
1. Helium—Popular works 2. Periodic law—Popular works.
I. Title. II. Series.
QD181.H4H37 2007
546'.751—dc22

2005029027

Manufactured in the United States of America

On the cover: Helium's square on the periodic table of elements.
Inset: A model of helium's subatomic structure.

Contents

Introduction

O h, the humanity!" was the cry of radio reporter Herbert Morrison as he watched the hydrogen-filled *Hindenburg*, carrying ninety-seven people, burst into flames. The *Hindenburg* was a magnificent airship built by Germany's Zeppelin Company in 1936. It was truly a marvel, measuring 804 feet (245 meters) in length and able to hold 6,710,000 cubic feet (190,006 cubic meters) of gas. Inside, it had a dining room, a writing room, and a lounge for the enjoyment of its passengers.

After making ten trips across the Atlantic Ocean in 1936, the *Hindenburg* was destroyed by fire in 1937 while landing in Lakehurst, New Jersey. It took only thirty-seven seconds for the hydrogen within the ship to burn, destroying the mighty *Hindenburg*. One-third of its crew and passengers were killed, and spectators were left crying in horror. Though the *Hindenburg* is remembered for its fiery crash and the deaths of thirty-six people, it was never meant to be filled with hydrogen at all. The designers of the *Hindenburg* had originally wanted to fill it with helium gas, not hydrogen. Unlike hydrogen, helium does not burn, making it safer to use for airships. The *Hindenburg* never got its helium, though.

In order to keep the company going during the Depression, executives at the Zeppelin Company had accepted large sums of money from the powerful National Socialist Party, better known as the Nazis. In fact, the majestic *Hindenburg* bore the swastika, the symbol of the Nazis, on its

The *Hindenburg* burst into flames just 200 feet (61 meters) from its intended landing spot on May 6, 1937. As fire engulfed the *Hindenburg*'s tail, many passengers began jumping. In the end, thirteen passengers, twenty-two on-flight employees, and one ground crewman lost their lives. Hydrogen, the gas that enabled the *Hindenburg* to float, is a highly flammable gas. Many believe that this tragedy could have been avoided if helium gas had been used instead.

fins. The United States controlled the helium supply and was becoming more and more suspicious of Adolf Hitler, the leader of Germany. When the United States, which controlled the manufacture of helium, refused to give the Zeppelin Company the helium it needed, the company was forced to fill the airship with flammable hydrogen gas instead. Unfortunately, this tragedy probably could have easily been prevented with the use of helium. Due in part to the lesson that was learned through the *Hindenburg* incident, today's airships are filled with nonflammable helium. Though hydrogen has greater lifting power than helium, helium is by far the safer choice.

Chapter One
Introduction to Helium

Helium (He) is an extraordinary gas. Though it is in the air around you, you would not know it. Like the other elements that make up air, you cannot see it, feel it, or even smell it. Unlike the other elements in the air, however, helium gas does not support life. In fact, it does not do very much at all. For this reason, it took scientists a very long time to discover it. When they did discover it, they found that it was very unusual. Because scientists could not get helium to react with any other chemicals, they thought that it was not a very useful element. However, now that chemists know more about helium, they have found many fascinating ways to use it. Helium is truly an amazing element!

Helium is the second most abundant element in the universe, after hydrogen (H). Astonishingly, helium and hydrogen together make up 99.9 percent of all the elements in the universe. However, most helium is found in outer space. Very little is actually found here on Earth. If you take a deep breath, you may inhale some helium, but not very much. Helium ranks a poor sixth among the gases that make up our atmosphere, comprising only about 0.0005 percent of it. Helium is not bound by gravity, and that's why there is such a small amount on Earth.

In space, helium is concentrated in stars, including our sun. Helium is formed in the hot stars by the fusion of hydrogen. Fusion is the merging

Helium makes up about a quarter of the mass of the sun. Helium in the sun is believed to be formed by nuclear fusion. In the sun, hydrogen (the lightest element) fuses together to make helium. This releases huge amounts of energy in the form of light and heat. Getting two positively charged protons to fuse together takes extreme heat and high densities. The only place these conditions occur naturally is in the center of stars, which includes our sun.

of two or more things. Scientists refer to the fusion of hydrogen into helium as "hydrogen burning." It is this process that provides the stars with the energy they need to shine. Without it, we would not be able to enjoy the warmth and light of the sun.

On Earth, nearly all helium is found trapped underground. In fact, 3,000 times more helium is found trapped below the surface of Earth than is found in Earth's atmosphere. Much of the helium that is found underground is located in reservoirs of natural gas. Natural gas is a mixture of gases trapped in the pores of underground rocks. Helium is formed underground when radioactive, or unstable, elements in these rocks decay, or break down. When elements, such as uranium (U) and radon (Rn) decay, they spontaneously give off helium. Some of this helium seeps up through cracks in the rocks and escapes into the atmosphere, while some remains trapped within the rocks.

The Discovery of Helium

In 1785, a man named Henry Cavendish (1731–1810) found that air was not made only of oxygen (O_2), nitrogen (N_2), and carbon dioxide (CO_2)—the only gases known at that time to be in the air. Unfortunately, this discovery was largely forgotten. Even Cavendish thought that his results could be explained by experimental error. However, helium would be discovered more than eighty years later by French astronomer Pierre-Jules-César Janssen (1824–1907). He discovered helium in 1868 while studying the sun using a spectroscope. A spectroscope is an instrument that identifies elements by the spectrum of lines they produce when they are heated. Each element, when heated, produces its own spectrum of colored lines. While examining the spectrum from the sun, Janssen noticed a yellow line that did not belong to any element that was known at that time. Meanwhile, British astronomer Sir J. Norman Lockyer (1836–1920) also observed the mysterious yellow line. Lockyer concluded that it belonged to an element that was common in the sun but was unknown on Earth. Therefore, he named the element helium, after Helios, the Greek god of the sun. Many scientists doubted the existence of helium. Even so, a search for the new element on planet Earth began.

The hunt to find helium on Earth finally ended in 1895. Sir William Ramsay (1852–1916), a Scottish chemist, discovered helium on Earth while experimenting with clevite, a uranium-containing mineral. A mineral is a crystalline substance of definite composition that occurs naturally

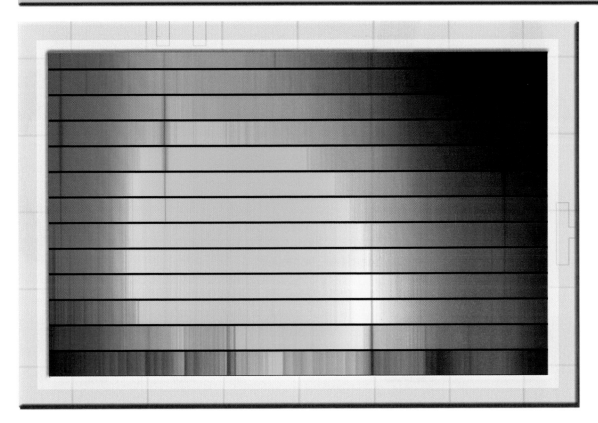

A star is a large ball of gas that creates and emits its own radiation, or light. A spectrum shows the distribution of colors emitted by an object. Stars are classified by their spectra and their temperature. Hotter stars put out most of their light in the blue region of the spectrum, while cooler stars put out most of their light in the red region. Here, the spectra of thirteen types of stars are being compared.

in rocks. Ramsay exposed the clevite to mineral acids and collected the gases that were produced. He sent a sample of the gases to Lockyer and Sir William Crookes (1832–1919), a British physicist. They were able to identify helium as one of the gases by using the same spectroscopic method used to identify helium in the sun. Two Swedish chemists, Nils Abraham Langlet (1868–1936) and Per Teodor Cleve (1840–1905), also independently discovered helium in clevite at about the same time as Ramsay. Helium is the only element to be identified in space before being found on Earth!

Helium He Snapshot

2 4
He

Chemical Symbol:	He
Classification:	**Nonmetal**
Properties:	**Nonreactive, nonflammable, colorless, odorless, nontoxic**
Discovered by:	**Pierre-Jules-César Janssen, in 1868, while studying the sun using a spectroscope. Helium was discovered on Earth by Sir William Ramsay, Nils Abraham Langlet, and Per Teodor Cleve in 1895.**
Atomic Number:	**2**
Atomic Weight (rounded):	**4.0026 atomic mass units (amu)**
Protons:	**2**
Electrons:	**2**
Neutrons:	**2**
Density at 68°F (20°C):	**0.0001664 gram/cubic centimeter (0.1664 g/liter)**
Melting Point:	**−457.87°F (−272.15°C)**
Boiling Point:	**−452.09°F (−268.94°C)**
Commonly Found:	**Air, stars, underground**

Dmitry Ivanovich Mendeleyev was born in Tobol'sk, Siberia, on February 7, 1834. His idea to organize the elements into a chart is the basis for the modern periodic table. Thanks to him, it is much easier for us to classify materials. Though helium was not known to Mendeleyev, he was very interested in human flight. He made an ascent in a balloon in 1887 to study a solar eclipse.

Helium and the Periodic Table

Everything on Earth, as well as throughout the universe, is made of one or more elements. Today, there are more than 100 known elements. As scientists discovered more and more elements over the years, they realized that there were similarities between the properties of some elements. Eventually, the elements were arranged on a big chart, called the periodic table. The periodic table that we use today is based on the work of a Russian chemist named Dmitry Mendeleyev (1834–1907). He published the first version of the periodic table in 1869, while teaching chemistry at the University of St. Petersburg in Russia. Mendeleyev sought to organize the elements in a way that would make studying and understanding them easier for his students. He arranged elements with similar properties into columns. Mendeleyev's periodic table did not list all of the elements that we know about today. Helium was unknown to him, and therefore was not included on his first chart. Helium was added to the periodic table sometime after its discovery on Earth in 1895.

Chapter Two
The Element Helium

Every element is made up of only one kind of atom. This means that every atom of helium is nearly the same. Atoms are very tiny. In fact, it would take 200 million helium atoms, lying side by side, to form a line only 0.4 inches (1 centimeter) long! Amazingly, atoms are made up of even smaller components called subatomic particles. In order to understand what truly makes helium atoms unique, we have to take a closer look at these particles.

There are three subatomic particles that make up the atom: neutrons, protons, and electrons. Neutrons and protons are clustered together at the center of the atom to form a dense core called the nucleus. Most of the mass of an atom comes from its neutrons and protons, and the mass of a proton is nearly the same as the mass of a neutron. Neutrons carry no electrical charge, while protons have a positive electrical charge. This gives the nucleus an overall positive electrical charge. Every helium atom has two protons in its nucleus, so its nucleus has a charge of +2.

Electrons are negatively charged particles that are arranged in layers, or shells, around the nucleus of an atom. The electrons are not fixed in a single position but circulate around the nucleus. The negative electrons are attracted to the positive nucleus, and it is this attraction that holds the electrons around the nucleus. An atom contains the same number of electrons and protons, so the negative and positive charges balance. Therefore, since a helium atom has two protons, it also has two electrons.

All Elements Are Unique

What makes helium different from other elements such as oxygen or silver (Ag)? The difference lies in the number of protons in the nuclei of the atoms. The number of protons in an atom's nucleus determines which element it is. This number is called the atomic number. On the periodic table, this number is found to the upper left of the element's symbol. In the periodic table, the elements are arranged in order of increasing atomic number. The atomic number of helium is two, which shows that an atom of helium has two protons in its nucleus. The fact that helium has two protons in its nucleus is what makes it helium. If you were able to add one proton to helium's nucleus you would end up with an entirely different element. Adding another proton would give you the element lithium (Li), which has three protons in its nucleus. Unlike helium, lithium is a metal. Just one proton makes the difference between a colorless gas and a bright, shiny metal. If you were to remove one of helium's protons, you would have hydrogen, which has only one proton in its nucleus. Hydrogen is very different from

Each helium atom has two protons and two neutrons in its nucleus. The protons carry a positive charge while the neutrons carry no charge. Two electrons revolve around a helium nucleus in one shell. These electrons carry a negative charge. It is the arrangement of these atomic particles that distinguishes helium from other elements.

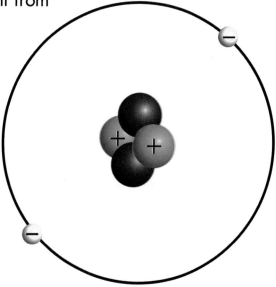

helium. While hydrogen will burst into flames when exposed to fire, helium will not burn under any circumstances. As you can see, by changing the number of protons, we get an entirely different element.

Atomic Weight

Cleve and Langlet, two of the chemists credited with the discovery of helium on Earth, were also the first to determine its atomic weight (also known as atomic mass) correctly. This number is found at the upper right-hand corner of an element's symbol on the periodic table. The atomic weight is the average weight of an atom of the element. It is measured in atomic mass units (amu). Atomic weight is an average weight and not an exact weight because elements often occur as mixtures of two or more isotopes. Isotopes are atoms with the same number of protons and electrons but different numbers of neutrons. The atomic weight is the average weight of all the element's isotopes, with consideration of how often each isotope occurs. Helium has two main isotopes and an exact atomic weight of 4.002602 amu, which rounds to four.

Arranging the Periodic Table

The periodic table that we use today lists the elements in order of increasing atomic number (the number of protons). By looking at the periodic table, you can see many trends, or patterns, in the properties of elements. You can use these trends to help you classify an element. By seeing where an element is located on the periodic table, you can predict whether it is a metal, a nonmetal, or a metalloid. Metalloids, or semimetals, have characteristics of both metals and nonmetals.

If you look at the periodic table, you will notice that the elements are divided by a "staircase" line. The metals are found to the left of this line, and the nonmetals to the right. Most of the elements bordering the line are metalloids. Helium is found to the right of the staircase line. This is where you would expect to find it, since helium is a gas, which means it is definitely not a metal. A nonmetal is an element that does not have the characteristics of a metal. Metals are easily recognized by their physical traits. Generally, metals can be polished to be made shiny. They also conduct electricity. Most metals also have the ability to be hammered into shapes without breaking. This is called malleability. Metals are also usually ductile. This means that they are able to be pulled into wires. Substances are classified as nonmetals if they lack the characteristics of metals. Nearly half of the nonmetals are colorless gases, but nonmetals can also be liquids and solids. Unlike solid metals, however, solid nonmetals are brittle and will crumble or break apart if pulled.

As you look across the periodic table from left to right, the horizontal row of elements is called a period. Elements are arranged in periods by the elements' atomic numbers. From an element's location on the periodic table, a person can determine the arrangement of the electrons in an atom. Helium is in period one, meaning it has only one shell of electrons surrounding its nucleus.

As you read down the periodic table from top to bottom, the column of elements is called a group or family. Just as you might have similar characteristics to the other members of your family, the elements in a given group have similar properties. Helium heads up the family of elements that make up group O, or 18. This group of elements is also referred to as the noble gases. The noble gas family consists of six gaseous elements: helium (He), neon (Ne), argon (Ar), krypton (Kr), xenon (Xe), and radon (Rn).

The Noble Gases

The electrons surrounding the nucleus of an atom are arranged in shells. Each of these shells is able to hold a certain number of electrons, and electrons normally fill the shells closest to the nucleus first. For instance, the

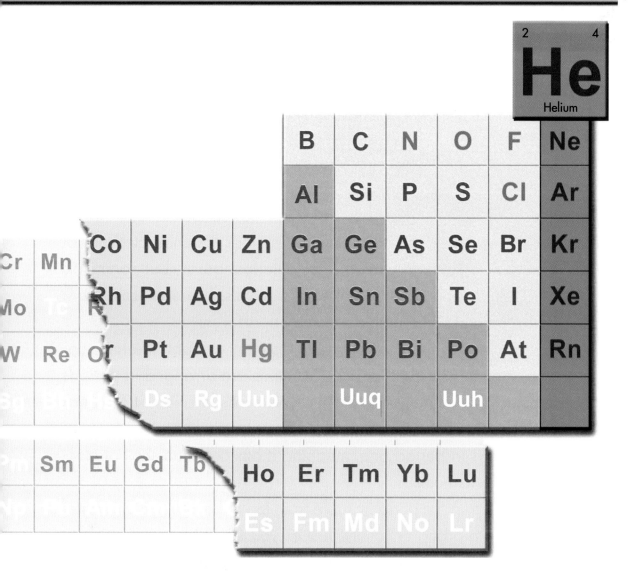

Helium heads the column known as the noble gases. They are called noble gases because they do not naturally react with other elements. In addition to helium, other noble gases include neon, argon, krypton, xenon, and radon. The noble gases are all colorless and odorless at room temperature. Their boiling points are all well below 32 degrees Fahrenheit (0 degrees Celsius), the freezing point of water.

shell closest to the nucleus of an atom is able to hold two electrons, while the second shell out is able to hold eight. The electrons in the outermost shell of an atom are called valence electrons. It is these electrons that determine how an element acts. The noble gases are different from all

The Hydrogen Bomb

The hydrogen bomb gets its energy from the nuclear fusion of hydrogen. Two lighter hydrogen nuclei combine to form a single, larger helium nucleus. The mass of the resulting helium nucleus is slightly less than the combined masses of the two hydrogen nuclei. The result is the release of a large amount of energy. The scientist Albert Einstein (1879–1955) formulated a special theory of relativity, $E=mc^2$, which explains mathematically how the missing mass relates to the energy that's released. The E stands for energy, the m represents mass, and the c is for the speed of light. Fusion reactions are also called thermonuclear reactions, so hydrogen bombs are called thermonuclear weapons. The first thermonuclear bomb was exploded as a test by the United States in 1952. Countries today have access to producing thermonuclear weapons, and some, like the United States, have done test explosions of them. However, presently, hydrogen bombs have never been used in war.

The first hydrogen bomb test was at Enewetok Atoll, Marshall Islands. The bomb was hundreds of times more powerful than the atomic bomb dropped on Hiroshima, Japan, during World War II. Whereas atomic bombs get their energy by splitting atoms, hydrogen bombs get theirs by forcing atoms together. This hydrogen bomb destroyed an entire island.

other elements because they have filled outer electron shells. For some atoms, such as those that belong to metals, the valence shell is nearly empty. For other elements, such as the nonmetals, the valence shell is nearly full. Therefore, elements seek either to donate, share, or accept electrons from other elements in order to fill them.

Atoms are very stable when the outermost electron shell is full. Since the noble gases have full outer shells, they are quite stable and rarely react with other elements. This explains why hydrogen burns but helium doesn't. Hydrogen atoms have a half-filled outermost shell, so hydrogen is a reactive element, but helium has a filled outer shell so it is nonreactive. For this reason, the name "noble gases" is very fitting for this group of elements. Like the kings and queens of nobility, the noble gases do not interact with other elements unless they are forced to do so.

Chapter Three
The Properties of Helium

All elements have characteristic physical and chemical properties. These properties help scientists to identify and classify them. The physical properties of an element are those that can be observed without changing the element's identity. Some examples of physical properties are an element's phase, or physical state, at room temperature; its density; and its boiling point. The chemical properties of an element describe the element's ability to undergo chemical change by combining with other elements. A chemical change converts one kind of matter into a new kind of matter. If an element undergoes chemical change easily, it is said to be reactive.

At room temperature, an element is found in one of three phases: solid, liquid, or gas. Knowing the phase of an element at room temperature helps scientists to identify it. Helium is found in the gas phase at room temperature. In fact, all of the noble gases are colorless, odorless gases at room temperature.

Gases, such as helium, do not have definite shapes or volumes. If put in a container, a gas will take the shape of that container. Gases differ from solids and liquids in that they are able to be compressed and are able to expand. A gas will fit in a container of almost any size and shape. If put in a large container, a gas will expand to fill it. A gas can also be

Ice   Water   Gas

Molecules in the solid and liquid phases touch each other. In the solid phase, molecules are "frozen" in place, making a regular arrangement. In the liquid phase, molecules are able to "flow" past each other. Molecules are much more spread out in the gas phase than they are in the solid and liquid phases. Because molecules in the gas phase are so far apart from each other, gases such as helium have a much greater volume than solids and liquids.

compressed to fit into a smaller container. If not confined to a container, gases will disperse into space.

Density is the mass of a substance contained in a specific volume. The density of a substance is found by dividing the mass by its volume. Each element has a unique density, and an element's density is useful in helping scientists to identify it. Helium has a density of 0.0001664 gram/cubic centimeter (0.1664 g/liter) at 68 degrees Fahrenheit (20 degrees Celsius). The density of different gases is often compared to the density of air. If a gas has a lower density than air, an object containing it will float in air. In contrast, if a gas has a higher density than air, an object containing it will sink in air. If you have ever seen a helium balloon floating, you know that helium has a lower density than air's density under the same conditions.

At normal room temperature, helium is always a gas. However, if it is cooled to a low enough temperature or squeezed under enough pressure, helium can become a liquid or even a solid. In order for helium to condense, or turn into a liquid, the temperature must reach a chilly −452.09°F (−268.94°C). This is the lowest condensation point (also called boiling point) of any known substance! In fact, this temperature is only 7.58°F (4.21°C) higher than absolute zero (−459.67°F [−273.15°C]), the lowest temperature possible.

Liquid helium is colorless and is so transparent, or see-through, that it is impossible to see its surface with the naked eye. For helium to reach its freezing point

Helium can be distributed as either a gas or a liquid. As a gas at normal temperature, helium is distributed in steel or aluminum alloy cylinders at high pressures. When it is necessary to distribute large quantities of helium, it is liquefied and transported in insulated, pressurized vessels, such as this one. Such containers can have capacities of up to about 14,800 gallons (56,024 liters).

Helium and Medicine

Magnetic resonance imaging (MRI) scanners have been saving lives since the 1980s with the help of helium. The machines use a very large magnet to produce detailed pictures of the inside of the human body. Doctors use MRI scanners to diagnose conditions such as brain tumors, eye abnormalities, and bone infections. MRI scanners use very powerful magnets that generate strong magnetic fields. In order to create a powerful enough magnetic field to produce scans, the magnet must be cooled. The extremely low temperature of liquid helium makes it an ideal coolant for MRI scanners. An MRI scan is harmless to the patient.

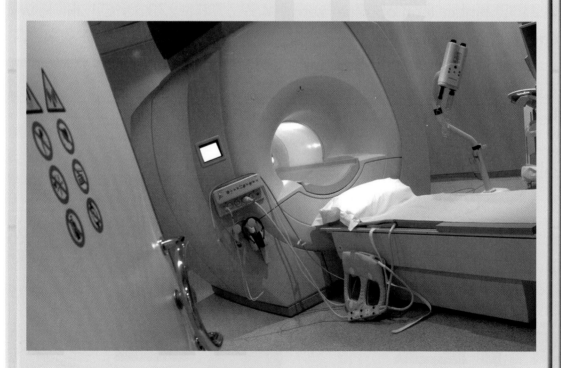

The first MRI scan ever to be performed on a human took place on July 3, 1977. As late as 1982, there were only a handful of MRI scanners. Today, there are thousands. A person being scanned by an MRI machine is slid into the magnetic tube on a special table. The MRI system creates a 2-D or 3-D map of the tissue types inside the patient's body.

(the temperature at which it becomes a solid), the temperature must drop even further to −457.87°F (−272.15°C). However, helium is very difficult to solidify. In fact, it is the only element that cannot be made into a solid at normal pressures simply by lowering its temperature. Scientists have only been able to solidify helium at extremely high pressures that are twenty-five times the normal atmospheric pressure.

The Fourth State of Matter

When helium becomes a liquid at −452.09°F (−268.94°C), it is called helium I. Helium I exists at just a few degrees above absolute zero. Helium I always bubbles, but if it is cooled to −455.76°F (−270.98°C), just a few degrees above absolute zero, it becomes perfectly still and begins to expand. At these low temperatures, liquid helium is called helium II, a fluid that behaves very strangely. It behaves so differently from other fluids that scientists often refer to it as a superfluid.

Viscosity is commonly measured in poise, a unit named after the French physiologist Jean-Léonard-Marie Poiseuille (1799–1869). Mercury (Hg), shown here, is the only metal that is a liquid at room temperature. Its viscosity at room temperature is 0.016 poise. This is close to water's viscosity at room temperature, 0.010 poise. It is difficult to imagine a substance that has nearly no viscosity at all, but that is the case with superfluids like helium II.

Superfluids are characterized by a lack of viscosity, and helium's lack of viscosity is the reason it's considered a superfluid. Viscosity describes how thick a liquid is or how much it resists being poured. For example, honey has a higher viscosity than water, which makes it thicker and harder to pour. At low temperatures, helium has almost no viscosity and feels no resistance, so it flows very easily. Because of this, helium II can fit through tiny holes that would be too narrow for most liquids to flow through. This also explains why, at these low temperatures, liquid helium will keep swirling once it is stirred. Helium II has even been known to flow uphill at low temperatures. If poured into a flask, it will first settle into the bottom, then it will start to climb up the sides, and soon it will flow out of the container!

Helium II can conduct heat 600 times as well as copper (Cu). This means that heat is able to move through it well. This is usually something that only solid metals can do well. In fact, of all the metals, copper is the best conductor of heat. The fact that the liquid, nonmetal helium is able to conduct heat so well is very strange indeed! Scientists find these properties so remarkable that some believe they have found a fourth state of matter.

Helium Isotopes

Helium atoms, like all those of the noble gases, exist in different forms called isotopes. Isotopes of atoms have the same number of protons and electrons but have a different number of neutrons. Because they have a different number of neutrons, they do not weigh the same amount.

Helium has two naturally occurring isotopes: helium-4 and helium-3. Helium-4 makes up more than 99 percent of all naturally occurring helium. Therefore, when people speak of "helium," they are usually referring to helium-4. Helium-4's nucleus consists of two protons and two neutrons for an atomic weight of four. The lighter isotope, helium-3, is

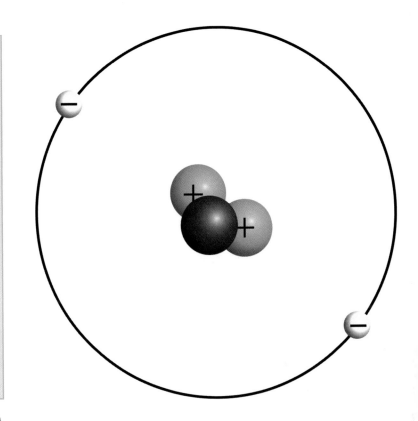

Helium-3, shown here, has one fewer neutron than ordinary helium (helium-4), which makes it the lighter isotope of the two. Both isotopes are stable. Unstable isotopes of helium have been synthesized, but since they are unstable, they cannot exist naturally. These unstable isotopes are helium-5, helium-6, and helium-8. Before the 1970s, it was believed that helium-3 could not become a superfluid. However, three scientists disproved that theory and won the Nobel Prize for their discovery.

much rarer. In Earth's atmosphere, there is one helium-3 for every million helium-4. The nucleus of helium-3 consists of two protons but only one neutron, for an atomic weight of three.

Different isotopes of the same element have almost identical chemical properties. This means that they react in the same way. However, helium-3 has a lower boiling point than helium-4 and exhibits different properties when liquefied. Like helium-4, helium-3 can also act like a superfluid, but that occurs at temperatures lower than those for helium-4.

Helium's Reactivity

As we have seen, helium and the other noble gases are very stable due to their full outermost electron shells. In fact, when the noble gases were first discovered, scientists could not get them to react with any other elements to form compounds. Compounds are substances that contain two or

When an electric current is passed through a gas, the gas emits light. The light is a result of collisions between atoms in the gas and electrons in the current. As electricity flows through this xenon-filled gas-discharge tube, it glows blue. When other noble gases are placed in the tube, they glow different colors. Neon emits a red glow, argon yields a violet glow, and helium produces a whitish orange glow.

more elements whose atoms have attached to each other in a characteristic arrangement. Because scientists thought the noble gases were unable to form compounds, the group was originally named the "inert gases." The word "inert" means nonreactive. In 1962, scientists were finally able to get one of the noble gases (xenon) to form a compound. Though helium is the least reactive member of the noble gases, it, too, is able to form compounds under extreme conditions. However, helium, like the other noble gases, must be forced to react. Helium can be made to form compounds with tungsten (W), iodine (I), fluorine (F), sulfur (S), and phosphorus (P). However, these compounds are very unstable, so they are unable to exist naturally.

Chapter Four
Collecting Helium

Helium is the second-lightest element, after hydrogen. Because it is so light, helium in Earth's atmosphere does not stay there very long but floats off into outer space. Because it would be expensive to extract helium from the air, the only way we can get the helium we need is to use our underground supplies. By far, the United States is the largest user of helium in the world, using about 3 billion cubic feet (85 million cubic meters) of helium each year.

Underground, helium is often found trapped in porous rocks that serve as reservoirs of oil and natural gas. Natural gas is the major commercial source of helium. Helium was first found in natural gas in 1903 by oil drillers in Dexter, Kansas. While drilling for oil, they created a gas geyser that contained an unknown gas. To their disappointment, this gas could not be burned for fuel. This gas turned out to be helium. Though helium had no uses at that time, it is very useful to us today.

Most of the world's supply of helium comes from underground gas wells in the United States. Great quantities of helium are found in Texas, Oklahoma, Arizona, and Kansas. Natural gas is a flammable mixture of gases. It consists mostly of methane (CH_4), which is used mainly for fuel. However, it also contains tiny amounts of other gases, such as helium.

Sir William Ramsay

Sir William Ramsay was born in Glasgow, Scotland, on October 2, 1852. Ramsay studied under the German analytical chemist Robert Bunsen at the University of Heidelberg. He went on to become a professor of chemistry at England's University of Bristol (from 1880 to 1887) and at the University of London (from 1887 to 1913). In addition to discovering helium on Earth, Ramsay is credited with having discovered four out of the other five noble gases. Radon, the last of the noble gases to be discovered, was not discovered until 1900 by a German scientist named Friedrich Ernst Dorn (1848–1916). Ramsay was known for spreading scientific interest to other parts of the British Empire. He even set up the Indian Institute for Science in Bangalore, India. Ramsay was elected fellow of the Royal Society in 1888, was knighted in 1902, and received the Nobel Prize for Chemistry in 1904. His

writings include *A System of Inorganic Chemistry* (1891); *The Gases of the Atmosphere* (1896); *Modern Chemistry*, 2 vols. (1900); *Introduction to the Study of Physical Chemistry* (1904); and *Elements and Electrons* (1913). Ramsay died on July 23, 1916.

Sir William Ramsay discovered helium while trying to find sources of argon in minerals. He also helped discover that helium is formed during radioactive decay of radium (Ra).

Getting Helium from the Ground

Drilling is used to locate natural gas reservoirs. The gas is then brought to the surface in pipes. The gas rises automatically because of the high pressure within the porous rocks where it is found. Once the gas has been extracted from the ground, it is usually transported in pipelines to a gas processing plant. Here, helium can be isolated from the rest of the gaseous mixture.

Isolating Helium

Helium is separated from methane and the other gases in natural gas by a process known as fractional distillation. Fractional distillation separates a mixture of liquids based on the boiling points of each element in the mixture. Every substance has a unique boiling point. As we have seen, helium's boiling point is −452.09°F (−268.94°C). Its boiling point is lower than that of all other substances. Because helium has such a low boiling point, it is very difficult to liquefy. Therefore, if natural gas is cooled, all of the other gases will turn into liquids first, leaving only helium as a gas. The helium that remains is then pumped off.

The natural gas, once it has been restored to normal temperature, is

Natural gas is mostly composed of methane, but it also contains concentrations as high as 4 percent of helium. Gas removed from the ground travels through pipelines like these to plants where the helium is extracted.

sold to gas companies. The helium is shipped out under pressures of 1,800 to 2,000 pounds (816 to 907 kilograms) per square inch in small containers or special tank cars. This helium is used for many different purposes, as you will see in the following chapter. The United States is the leading producer of helium (it produces 80 percent), followed by Algeria, Russia, Canada, Poland, and Qatar. The helium that is produced is about 99.995 percent pure. That is a pretty remarkable achievement!

Natural gas is separated from its components in a process called fractional distillation. First, all impurities that might freeze during the very cold process, such as carbon dioxide and heavy hydrocarbons, are removed in the preheating process. An expansion valve and a high-pressure fractionating column then cool the gas. The now-liquid methane goes through a low-pressure fractionating column, which removes more nitrogen. It is then pumped off and warmed and becomes an upgraded natural gas. The gases from the top of the high-pressure fractionating column are cooled in a condenser. This condenses most of the nitrogen, leaving only helium in a gaseous state. Once helium has been separated from the natural gas, it undergoes a purification process that will make it 99.9 percent pure and ready for commercial use.

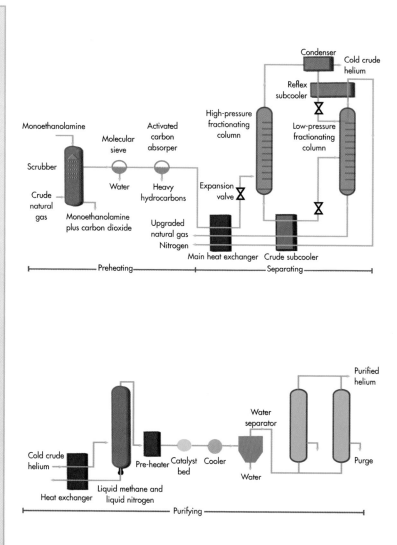

Breathing Helium

Most people know that breathing in helium makes a person's voice sound like a chipmunk's, but not everyone knows why. The temporary change in the voice of someone who has inhaled helium is related to the fact that helium is significantly less dense than air. As a result, the speed of sound in helium is nearly three times what it is in air. At 68°F (20°C), the speed of sound in air is 344 miles per second (554 kilometers per second) as compared to 927 m/s (1,492 km/s) in helium. Though helium is not toxic, it takes the place of oxygen in the lungs, which makes inhaling it dangerous if done in excess. This lack of oxygen could cause a person to pass out or even die.

Running Out of Helium

Helium is found in great quantities only in natural gas, and many scientists warn that supplies of helium could run out if they continue to be used so quickly. Many feel that once our current supplies of helium, found trapped underground, are used, we will be out of luck. In fact, some natural gas fields are nearing exhaustion. No shortage is envisioned in the near future, but many feel that it is sure to come eventually.

Running out of helium would mean the end of squeaky voices, but there are also more critical uses for helium. In fact, for many of its uses, there is no substitute for helium. There is no other substance that can take helium's place in circumstances in which temperatures below −429°F (−256°C) are needed. Though hydrogen could be substituted for helium in some lighter-than-air applications, such as in weather or research balloons, helium's advantage of being nonflammable would be lost. That can be a problem, as was evidenced by the *Hindenburg* disaster.

Chapter Five
Helium and You

Most people are familiar with the element helium, but not many know just how much it affects them. Because of its unique properties, helium has some pretty interesting uses. Helium has a low boiling point, it is unreactive, and it has a low density. Because of these qualities, helium is able to be used, among other things, in deep-sea breathing systems, to cool powerful magnets, to inflate balloons, to provide lift for airships, and as a protective gas in welding.

When divers use scuba, a self-contained underwater breathing apparatus, they breathe from a tank that holds highly compressed air. Water is much more dense than air, so it places a lot of pressure on divers' bodies, especially their lungs. A diver's lungs can be under so much pressure that they could collapse. To keep this from happening, the air coming out of a scuba tank is kept at the same pressure as the water around the diver. However, when high-pressure gases come in contact with water, they dissolve into the water. This is how carbonated beverages, such as soda, are made. To make carbonated water, the water is exposed to high-pressure carbon dioxide gas. This forces the gas to dissolve in the water. In the same way, a diver who stays in deep water for a while will have nitrogen and oxygen gases from the air tank forced into his or her bloodstream.

Excess nitrogen in the blood can cause a serious problem called nitrogen narcosis. This causes divers to feel as though they are drunk. A diver

In order to stay underwater for long periods of time, divers use breathing apparatuses that provide them with air. Helium is used to replace part of the nitrogen and oxygen in the breathing tanks of deep-sea divers. This idea was first proposed in 1919 by Elihu Thompson, an electronics engineer and inventor. Using a breathing mixture containing helium helps divers to remember the diving procedures, to perform better under stress, and to feel better after the dive.

suffering from nitrogen narcosis may have bad coordination and may make poor decisions. Many times, divers under this condition will throw their masks away and swim quickly into deeper water. These divers may not even be aware that their judgment is impaired. Other effects of nitrogen narcosis include numb lips, mouths, and fingers, or extreme exhaustion. Divers may panic or remain at the bottom, too tired to come up. Excess oxygen in the blood can also be a problem. Too much oxygen in the blood can cause oxygen toxicity. Oxygen toxicity results in convulsions, or uncontrollable shaking. This can cause a diver to drown.

Helium can help divers to avoid dangerous problems such as nitrogen narcosis and oxygen toxicity. Helium is not very soluble in water or blood. This means that not a lot of helium will enter a diver's bloodstream, even under high pressure. For this reason, in many breathing gases, part of the nitrogen and oxygen is replaced by helium. Helium is easy to breathe at high pressures and, unlike oxygen and nitrogen, it does not become toxic. Some common helium-containing breathing gases are trimix, a mixture of helium, oxygen, and nitrogen; heliox, which contains helium and oxygen; and hydreliox, a mixture of hydrogen, helium, and oxygen. These helium-containing breathing gases can help save a diver's life, but they can also make that diver sound a little bit like the Disney cartoon character Donald Duck.

Helium Preserving History

Two of the United States' most important documents are the Declaration of Independence and the Constitution. The Declaration of Independence marked the first official step toward separating the thirteen colonies from the control of Great Britain—and the first step toward creating the United States of America. The document was written by Thomas Jefferson (1743–1826). Though the Declaration of Independence was not officially signed until August 1, 1776, Congress voted to approve it on July 4, 1776. This is why the Fourth of July is celebrated as Independence Day today. Following the Revolutionary War (1775–1783), in which Americans fought the British to win their freedom and independence, the Founding Fathers crafted the Constitution. It was accepted by the states on May 29, 1790. The Constitution is the basis for the government of the United States. It explains how the government is formed, who makes it up, and how laws are to be passed and amended.

The Declaration of Independence and the Constitution were described by the poet and librarian of Congress Archibald MacLeish (1892–1982)

A guard stands next to the U.S. Constitution in the rotunda of the National Archives in Washington, D.C. The Constitution not only needs to be protected from attack by elements in the air, it must also be protected from natural and man-made disasters. At the first sign of vandalism, fire, or nuclear war, a device will plunge the document 20 feet (6 m) down into a fireproof, bombproof vault made of steel and reinforced concrete.

as "fragile objects which bear so great a weight of meaning to our people." Indeed, these priceless documents are quite fragile. They are handwritten with iron-gall ink on animal skin parchment. If left out in the air, atmospheric gases, such as oxygen, would surely destroy them. For this reason, it was decided that the documents would be enclosed in a case with the least reactive gas. What better than helium? In 1952, the frames holding the Declaration of Independence and the Constitution were emptied of air and then filled with 1 percent water vapor and 99 percent helium and placed in the National Archives in

Washington, D.C. What an honor for helium to be entrusted with such important documents!

Due to deterioration of the glass that enclosed the documents, they were transferred to new containers made of aluminum (Al), titanium (Ti), and glass in 2001. The enclosures that now surround the documents are filled with argon, another of the noble gases. Argon atoms are larger than helium atoms and are less likely to leak out. However, for nearly fifty years, helium did its job well.

Helium and Welding

Helium is very useful when it comes to welding. Welding is a process in which two or more pieces of metal are melted and joined together. Helium is used for both arc welding and laser welding. In arc welding, the heat that joins the metals together comes from electricity. Laser welding uses a laser beam to generate heat. Welding can be very dangerous. When the metals get very hot, they can react with oxygen in the air and start to burn. Oxygen can also ruin the metals that welders are trying to join together. For this reason, helium is used to protect the metal. Helium does this by "shielding" the metals from oxygen in the atmosphere. Helium is used for the welding of metals like stainless steel, aluminum, copper, nickel (Ni), titanium, and magnesium (Mg). As these metals are welded, helium gas is continuously blown over the area. Helium pushes the air away from the metals being welded, keeping oxygen and moisture from attacking them. Since helium is so unreactive, it does not react with the air

A welder wears a protective visor to shield him from hot sparks as he joins pieces of metal together. The heat created during welding can be potentially dangerous to both the welder and the metal he is welding. Helium is very useful in this situation. Helium can be used when an unreactive atmosphere is needed. Because it is unreactive, helium can be used to push oxygen-containing air away from the work area.

or the metals. Helium is also used in gas mixtures with other shielding gases, such as argon. Welding is used in the construction of bridges, ships, cars, and airplanes, among other objects. The next time you take a ride in a car or an airplane, think of helium's use as a welding agent.

It's Party Time

Helium balloons are used for everything from scientific research to military observation. However, everyone's favorite balloons are probably the colorful ones we see at birthday parties and circuses. Helium gained its fame as a lifting gas. It is lighter than air, so it floats on air in the same way that a plastic bottle filled with air floats on water.

Any gas that is less dense than air could actually be used to lift a balloon. Hydrogen, methane, ammonia (NH_3), and natural gas

A balloon is made of strong, light material that rises when it is filled with a gas that is lighter than air. Rubber balloons were first made in 1824 by a professor named Michael Faraday (1791–1867). He used them in experiments with hydrogen gas. Metallized nylon balloons first appeared in New York City in the 1970s. Often mistakenly called Mylar, metallized nylon balloons are made of two sheets of plastic (polyethylene) and nylon that are sandwiched together and coated with a thin layer of aluminum.

are all less dense than air, but they are all flammable. Early toy balloons were actually filled with hydrogen. However, they were not very safe. Even the slightest spark could cause a huge explosion. As early as 1914, concerned firefighters were trying to get hydrogen in toy balloons outlawed. In 1922, New York City banned hydrogen-filled toy balloons following a prank in which someone exploded hydrogen-filled balloon decorations at a city function, badly injuring an official. Though hydrogen is lighter than helium, hydrogen-filled toy balloons were eventually replaced by safer helium-filled ones.

Helium-filled rubber balloons deflate much more quickly than silvery metallized nylon balloons. This is because the rubber balloons have tiny pores through which the helium can escape. The aluminum coating of metallized nylon balloons does not have pores, so the helium stays inside longer, and the balloon stays up longer.

Toy balloons are very popular even today. Balloons are manufactured by the millions every day in a number of countries. They wind up at parties, carnivals, fairs, and circuses—adding a splash of color and a burst of excitement. Though helium was, at first, thought to be useless, it has definitely proven otherwise. This unique element has found a place in health care, helping us to diagnose illness. It serves to make diving, welding, and airship travel safer. It even helps to add a little more fun to parties. The world is definitely a better place because of helium!

The Periodic Table of Elements

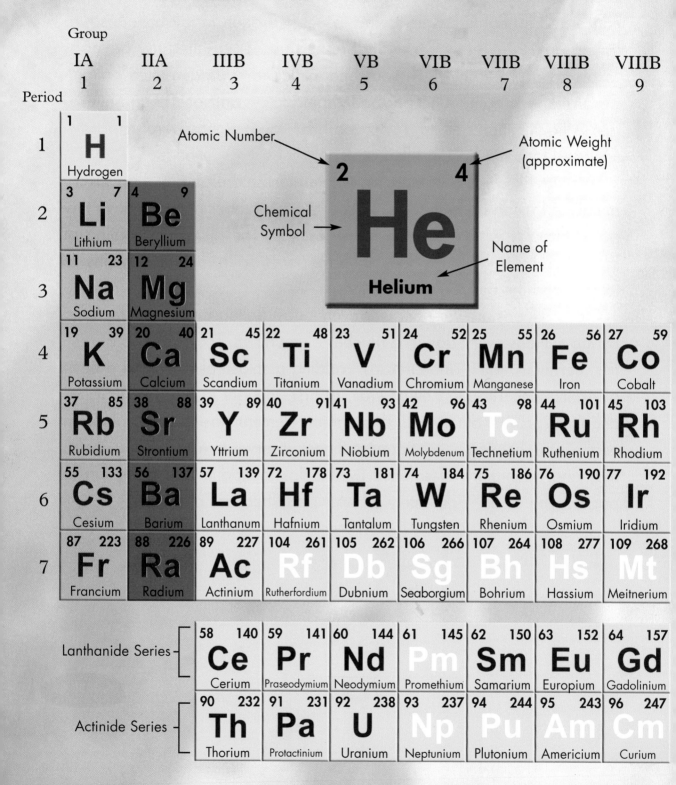

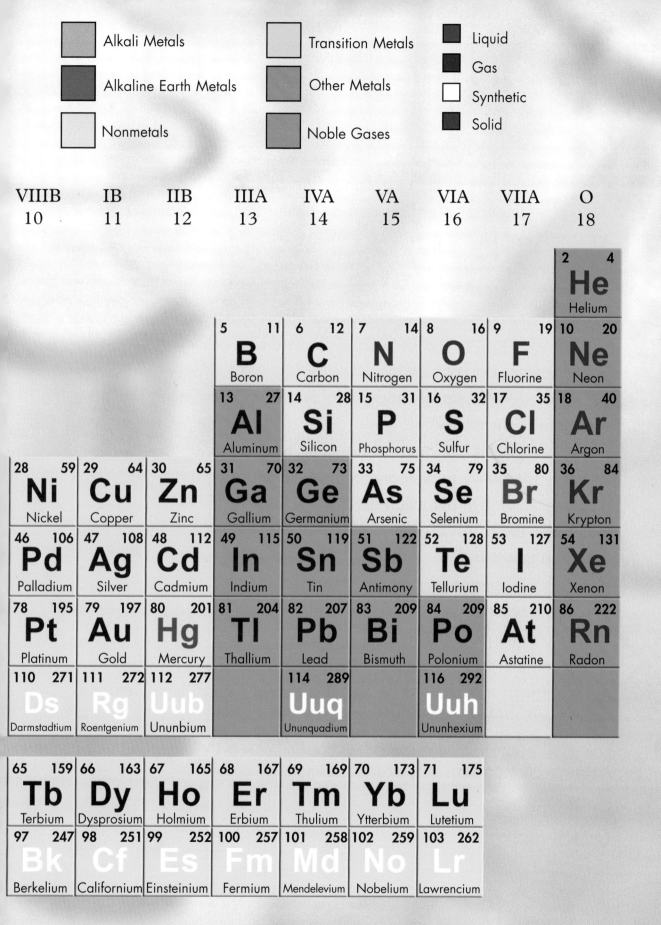

Glossary

atmosphere The air that surrounds Earth.

atom The smallest part of an element having the chemical properties of that element.

atomic weight The average mass (in units) of the atoms of the isotopes of an element.

bond An attractive force that links two atoms together.

chemical change A change in which one kind of matter is turned into another kind of matter.

compound A substance made up of two or more elements that have combined chemically.

element A substance made of only one type of atom.

energy The ability to do work (or produce heat).

gravity A force of attraction between masses.

magnetic field An area around a magnet where the magnetic force can be detected.

mass A measure of the amount of matter in something.

matter What things are made of. Matter takes up space and has mass.

pressure The force acting on a surface.

room temperature The average indoor temperature at which experiments are performed. Scientists consider room temperature to be approximately 68°F (20°C).

viscosity A fluid's resistance to flow.

volume The amount of space that something occupies.

For More Information

Compressed Gas Association
777 East Eisenhower Parkway
Ann Arbor, MI 48108
(800) 699-9277
Web site: http://www.cssinfo.com/info/cga.html

The Lighter-Than-Air Society and the Akron Airship Historical Center
526 S. Main Street, 232
Akron, OH 44311
Web site: http://www.blimpinfo.com

Web Sites

Due to the changing nature of Internet links, the Rosen Publishing Group, Inc., has developed an online list of Web sites related to the subject of this book. This site is updated regularly. Please use this link to access the list:

http://www.rosenlinks.com/uept/heli

For Further Reading

Asimov, Isaac. *The Noble Gases.* New York, NY: Basic Books, 1966.

Claassen, Howard H. *The Noble Gases.* Boston, MA: DC Heath & Co., 1966.

Earnshaw, Tim. *Helium.* Frankfurt, Germany: Eichborn, 1999.

Greenwood, N. N., and A. Earnshaw. *Chemistry of the Elements.* Oxford, England: Pergamon Press, 1984.

Hudson, John. *The History of Chemistry.* New York, NY: Routledge, Chapman, and Hall, 1992.

Vollhardt, Dieter, and Peter Woelfle. *The Superfluid Phases of Helium 3.* Boca Raton, FL: CRC Press, 1990.

Bibliography

Brady, James E., Joel Russell, and John R. Holum. *Chemistry: The Study of Matter and Its Changes.* New York, NY: John Wiley & Sons, Inc., 1993.

Ebbing, Darrell D. *General Chemistry.* 4th Edition. Boston, MA: Houghton Mifflin Company, 1993.

Stwertka, Albert. *A Guide to the Elements.* 2nd Edition. New York, NY: Oxford University Press, 2002.

Thomas, Jens. *The Elements: The Noble Gases.* New York, NY: Benchmark Books, 2002.

Index

About the Author

Heather Elizabeth Hasan graduated from college *summa cum laude* with a dual major in biochemistry and chemistry. She has written numerous books for the Rosen Publishing Group, including *Understanding the Elements of the Periodic Table: Iron* and *Understanding the Elements of the Periodic Table: Nitrogen*. She currently lives in Greencastle, Pennsylvania, with her husband, Omar, and their sons, Samuel and Matthew.

Photo Credits

T1. W2.

U3. X4.